How to use this book

A sample page

INSTRUCTION
What your child needs to do for the activity.

TITLE
The page title describes the skill your child will learn in these pages.

FUN ILLUSTRATIONS
Specially drawn illustrations which are fun, interesting and drawn at the right pedagogical level for your child.

COLOURFUL BORDERS
The page borders make each page as attractive as possible to stimulate your child.

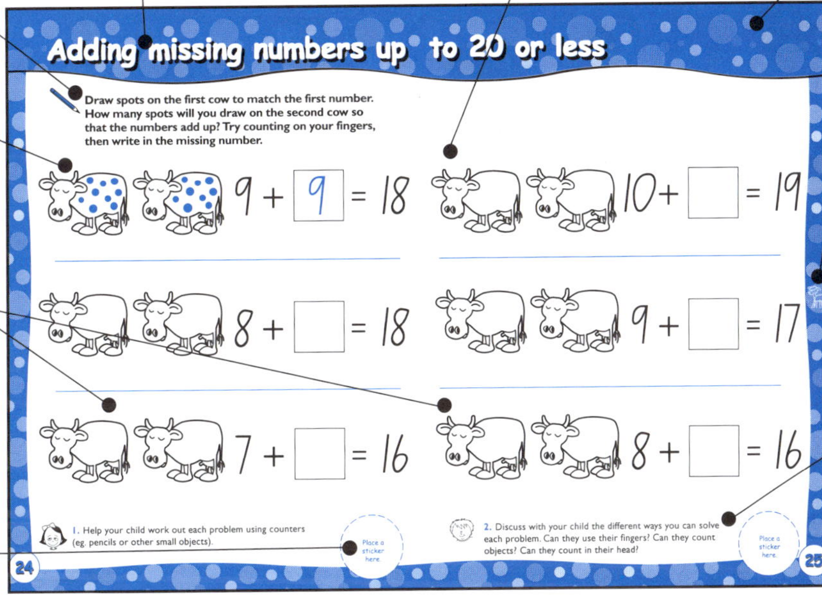

EXAMPLE
The first one is done for you so you can show your child exactly what to do.

LOTS OF PRACTICE
Two pages where your child can practise and repeat the same skill to master it.

STICKERS
Place a sticker on each page as your child finishes.

WHO'S HIDING?
In each book, a little creature appears in the border of every double page so your child can have fun trying to find it.

EXTRA ACTIVITIES
Extra activities you might want to do with your child to further reinforce the skill or simply make it more enjoyable.

Step-by-step learning

STEP ONE — **Read** out the title of the activity page to your child.

STEP TWO — **Explain** the skill and show your child the example already done. **Make sure** they understand what to do. Your child will then have at least two pages to practise that same skill.

STEP THREE — **Help** your child put a **sticker** on the bottom of each page as they complete it.

Remember to be patient, encouraging and positive with your child, even when minor mistakes are made!

How to hold a pencil

It is important that you help your child hold their crayon or pencil in the correct way as shown here to ensure your child develops the right technique early on.

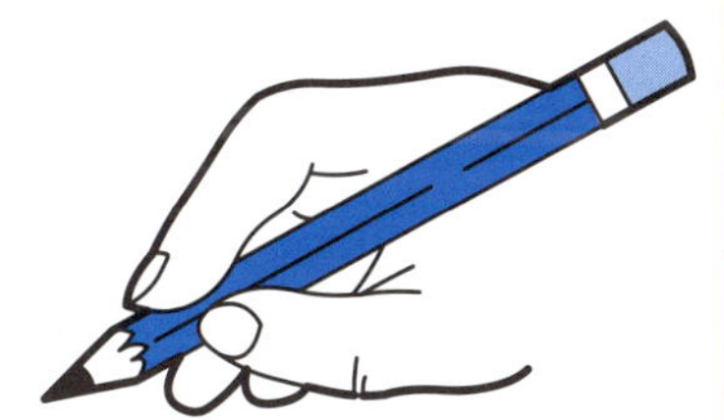

Adding up 10 or less

Draw fish in the two bowls to match the number story. Then count how many fish there are altogether and write the number in the circle.

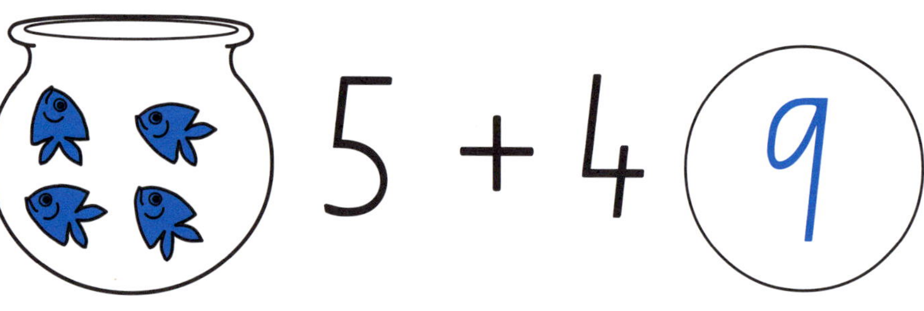

1. Ask your child to say each number story out loud. For example, '5 fish and 4 more makes 9 fish'.

Place a sticker here.

3 + 5

1 + 4

6 + 3

2. Play an adding game. You and your child each hold up some fingers on one hand. Ask your child to count how many altogether.

Place a sticker here.

Adding up to 10

What numbers add together to make 10? Take a red and a blue pencil. Colour squares in red to match the first number. Then colour the rest of the squares blue. Count the blue squares and write the number on the square card.

3 + [7] = 10

8 + [ ] = 10

1 + [ ] = 10

4 + [ ] = 10

1. Ask your child to say each number story out loud. For example, '3 and 7 makes 10'.

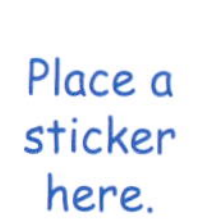

5 + ☐ = 10

2 + ☐ = 10

6 + ☐ = 10

10 + ☐ = 10

2. Give your child a group of 10 spoons or other items. How many different ways can they divide the spoons into 2 groups?

Place a sticker here.

Counting on to 10

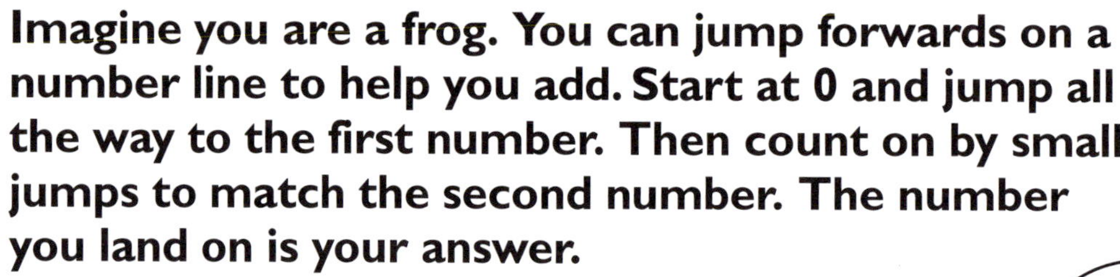

Imagine you are a frog. You can jump forwards on a number line to help you add. Start at 0 and jump all the way to the first number. Then count on by small jumps to match the second number. The number you land on is your answer.

$3 + 2 = 5$

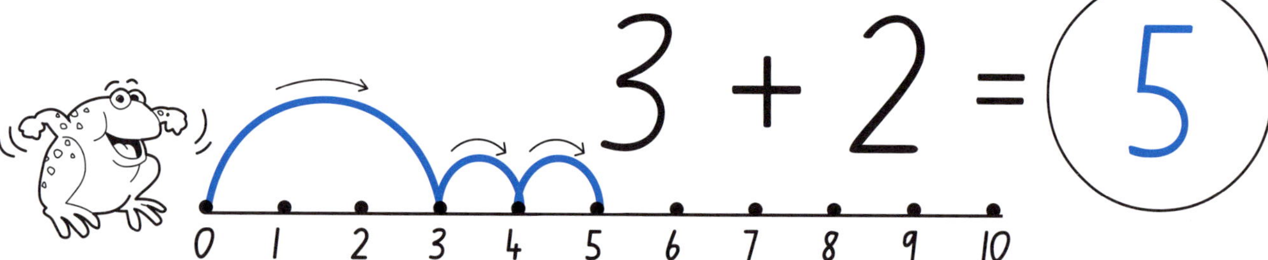

$4 + 3 =$

0 1 2 3 4 5 6 7 8 9 10

$5 + 1 =$

0 1 2 3 4 5 6 7 8 9 10

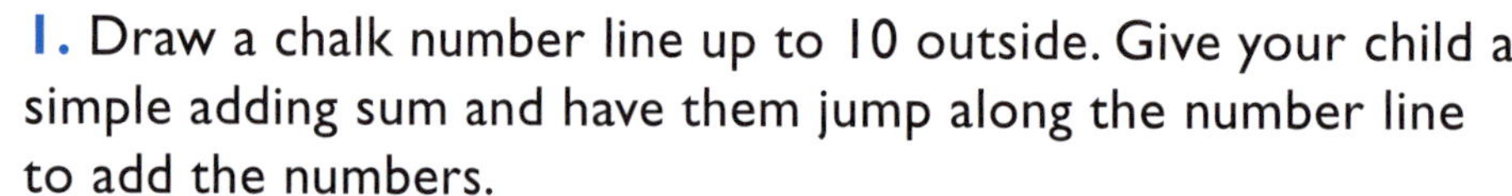

1. Draw a chalk number line up to 10 outside. Give your child a simple adding sum and have them jump along the number line to add the numbers.

Place a sticker here.

$4 + 4 =$ ◯

0 1 2 3 4 5 6 7 8 9 10

$5 + 4 =$ ◯

0 1 2 3 4 5 6 7 8 9 10

$7 + 3 =$ ◯

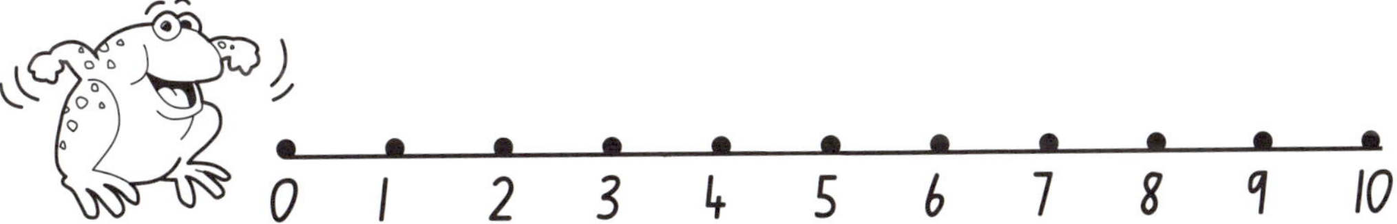

2. Repeat each addition on this page using buttons. Put buttons in a container matching the first number in the sum, then add more and count the total.

Place a sticker here.

Finding the partner to 10

The numbers on the shirts add to the number in the square. Count in your head or use your fingers to find the missing number. Then write it on the blank shirt.

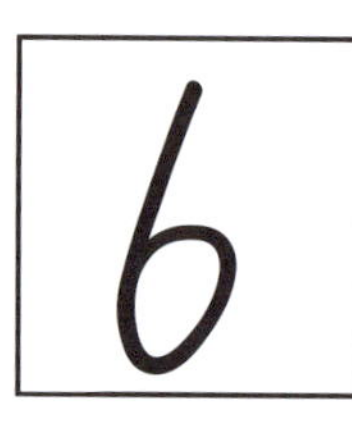

1. Draw pictures on a sheet of paper to match each sum on this page. For example, 4 balls and 3 balls makes 7 balls.

Place a sticker here.

8

9

6

8

4

9

2. Collect pieces of fruit to match one of the numbers in the squares. Divide them into two groups to find different addition combinations.

Place a sticker here.

Adding on from 10

Colour the beans in the pod to match the first number, then the loose beans to match the second number. Count up how many coloured beans there are altogether, then write the number in the boxes.

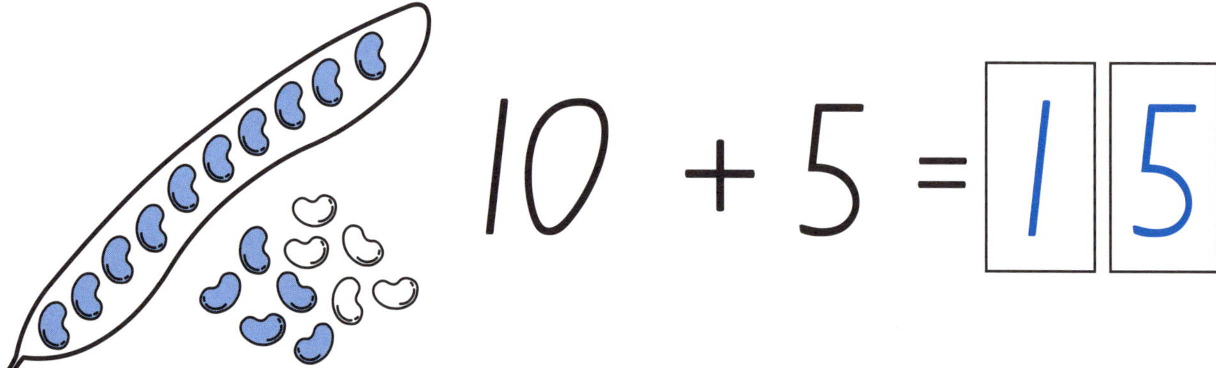

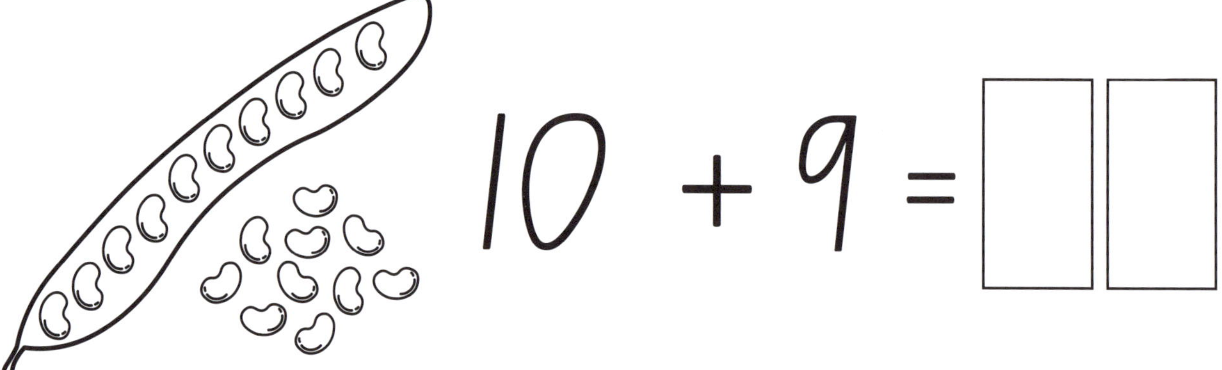

1. Make sure your child colours in the 10 beans on the pod first, then counts and colours in the extras.

Place a sticker here.

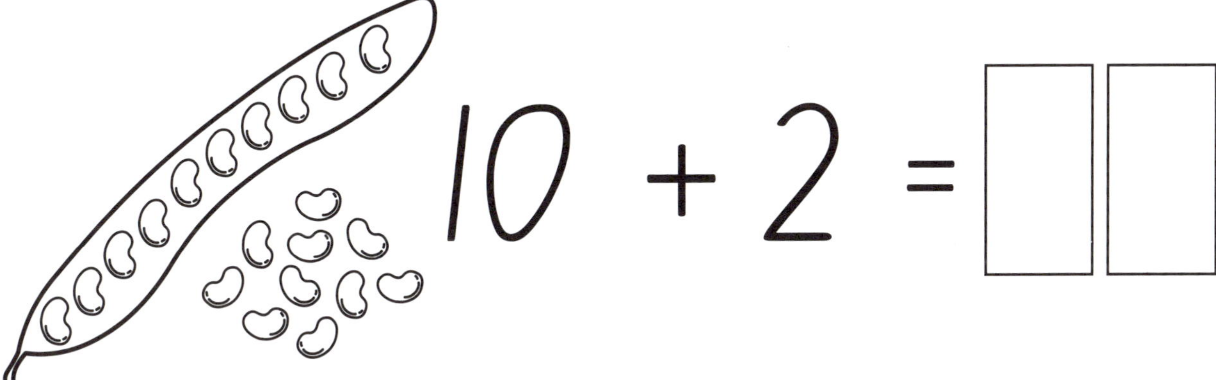

$10 + 2 =$ ☐☐

$10 + 10 =$ ☐☐

$10 + 4 =$ ☐☐

2. You can make your own 'beansticks' by pasting 10 dried beans on an ice-block stick and counting extra dried beans up to 20.

Place a sticker here.

Adding 10 to 0–9

Take a red and a blue pencil. Colour squares in red to match the first number. Then colour squares in blue to match the second number. How many squares are coloured altogether? Write the number in the boxes.

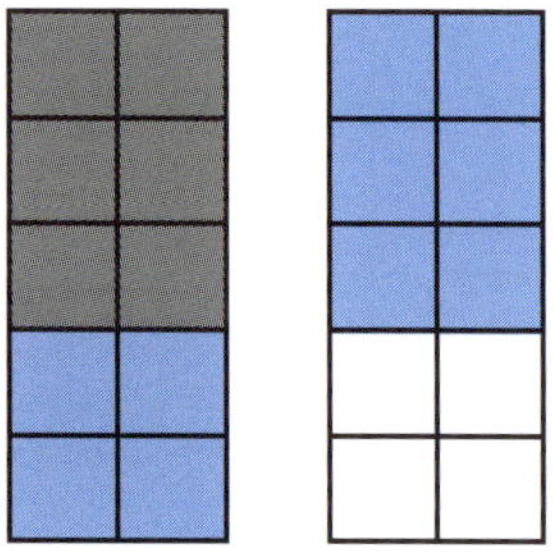

6 + 10 = 16

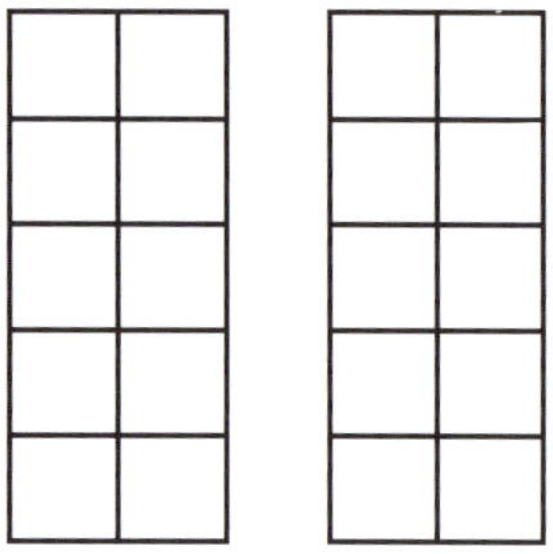

3 + 10 = ☐☐

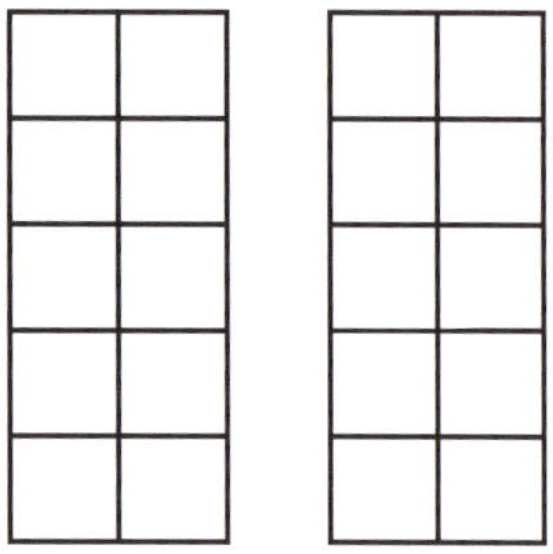

5 + 10 = ☐☐

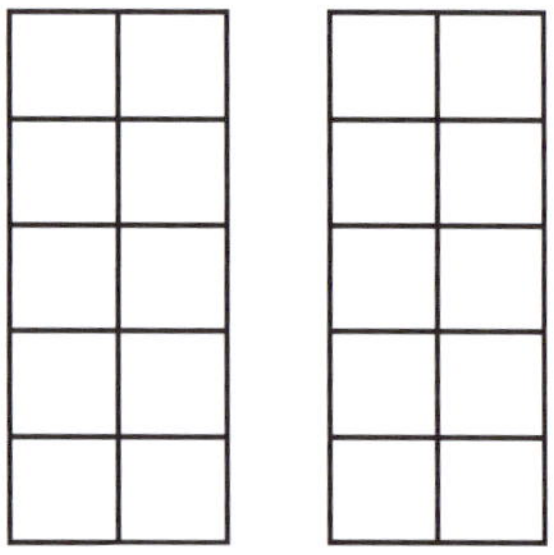

2 + 10 = ☐☐

1. Make sure your child colours all the squares in the first grid before colouring the extras in the next grid.

Place a sticker here.

9 + 10 = ☐☐

4 + 10 = ☐☐

7 + 10 = ☐☐

10 + 10 = ☐☐

2. Flash up to 20 fingers (eg. 5 fingers then 10 fingers) in front of your child for an instant. Can your child tell you how many fingers you held up? Try it again with a different number.

Place a sticker here.

Adding doubles

Draw spots on the first wing of each butterfly to match the first number. Then draw spots on the second wing to match the second number. How many spots are there altogether? Write the number in the boxes.

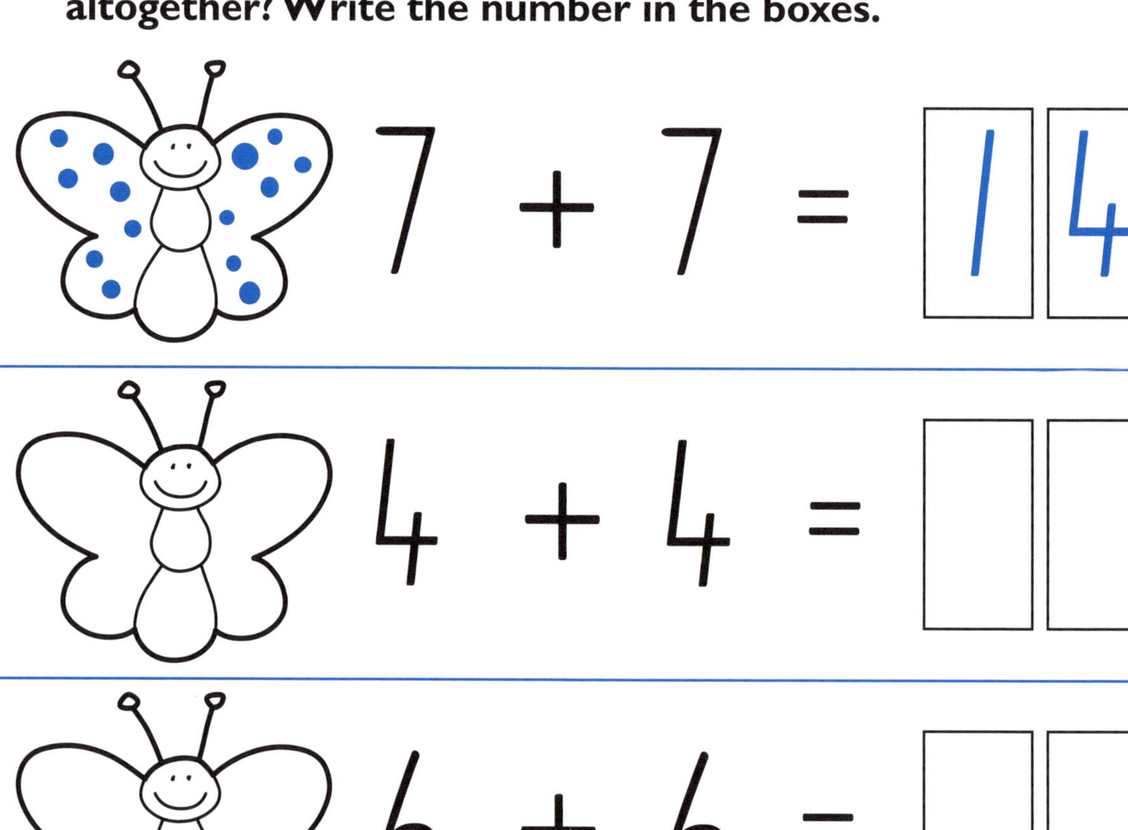

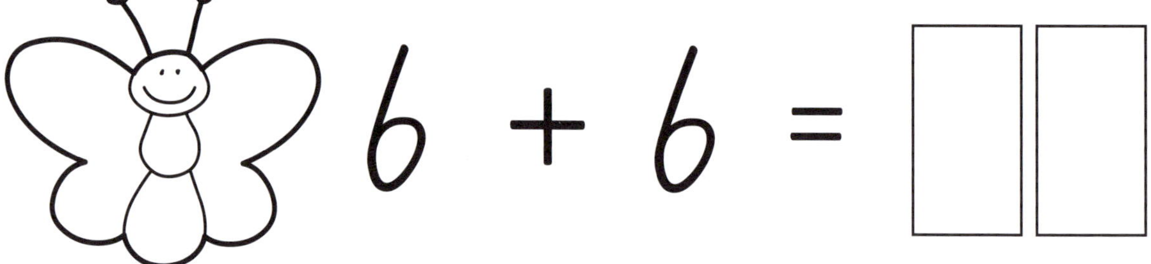

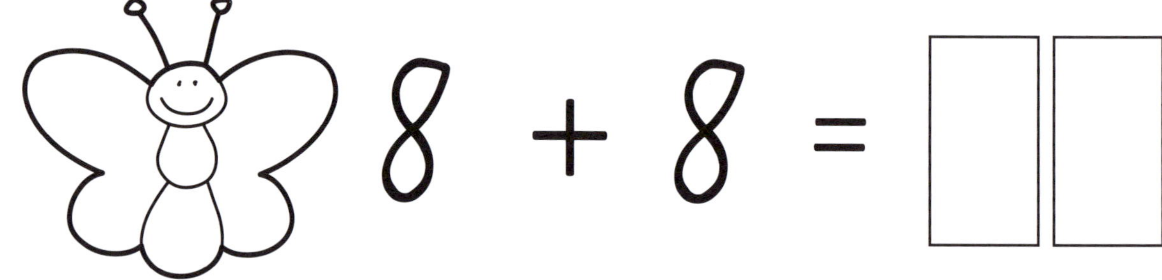

1. Throw two dice. Ask your child to clap their hands every time you throw a double. What numbers do the pairs add up to?

Place a sticker here.

9 + 9 = ☐☐

5 + 5 = ☐☐

8 + 8 = ☐☐

10 + 10 = ☐☐

2. Give your child a group of small counters. Call out a number between 1 and 10, and ask your child to arrange the counters into 2 groups of that number. How many are there altogether?

Place a sticker here.

Adding up to 15 or less

Draw apples on the first tree to match the first number. Then draw apples on the second tree to match the second number. How many apples are there altogether? Write the number in the boxes.

1. Throw two dice and ask your child to add the numbers. Do they need to count the dots or can they do it in their head?

Place a sticker here.

$9 + 3 =$ ☐☐

$6 + 8 =$ ☐☐

$9 + 6 =$ ☐☐

2. Hold up between 5 and 10 fingers and ask your child to count them. Now flash 5 more. Ask your child to tell you how many there are altogether.

Place a sticker here.

Adding missing numbers up

Draw hairs on the first boy's head to match the first number. How many hairs will you draw on the second boy's head so that the numbers add up? Try counting on your fingers, then write in the missing number.

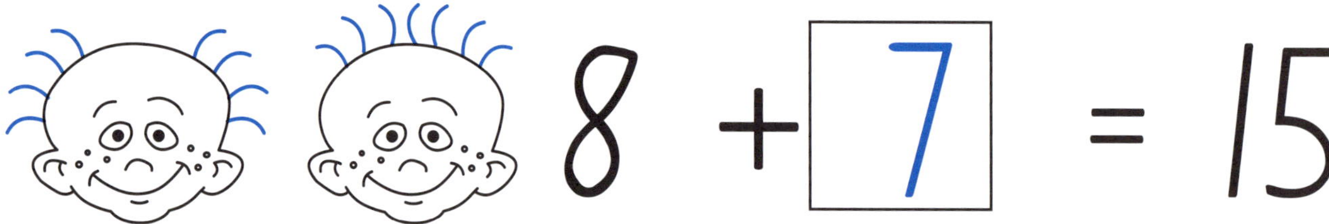

8 + [7] = 15

9 + [ ] = 14

7 + [ ] = 12

5 + [ ] = 13

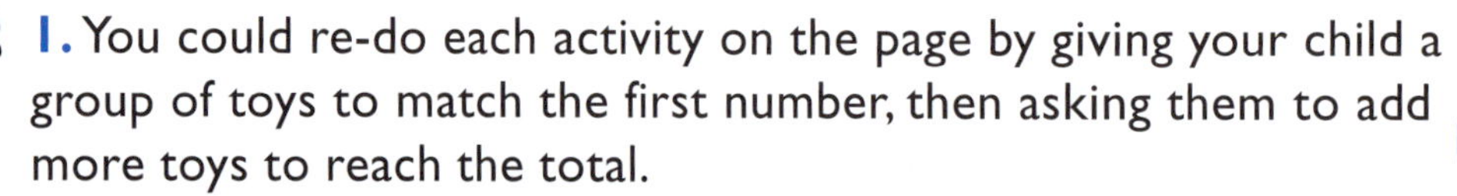

1. You could re-do each activity on the page by giving your child a group of toys to match the first number, then asking them to add more toys to reach the total.

Place a sticker here.

to 15 or less

$6 + \square = 13$

$4 + \square = 12$

$7 + \square = 14$

$10 + \square = 12$

2. If your child is confident with this activity, ask them if they can solve the problems in their head.

Place a sticker here.

Adding more missing numbers

Draw olives on the second pizza to match the second number. How many olives will you draw on the first pizza so that the numbers add up? Try counting on your fingers, then write in the missing number.

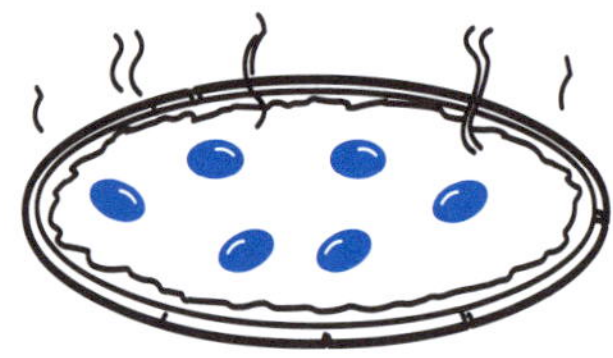 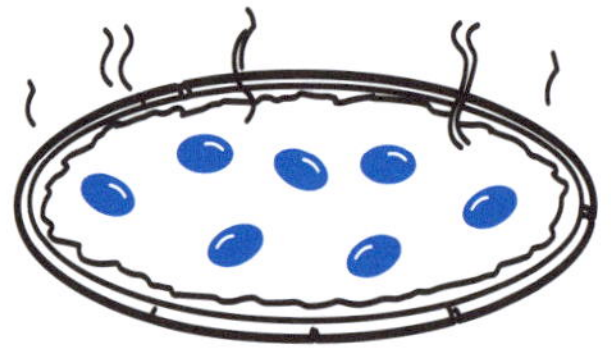 

+ 7 = 13

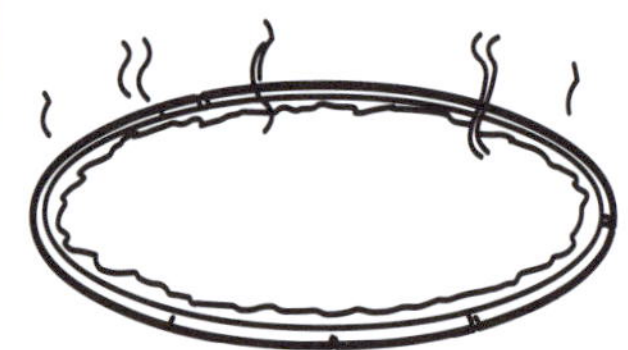 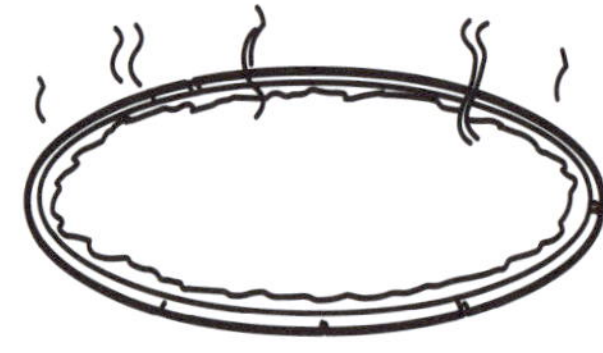 ☐ + 3 = 11

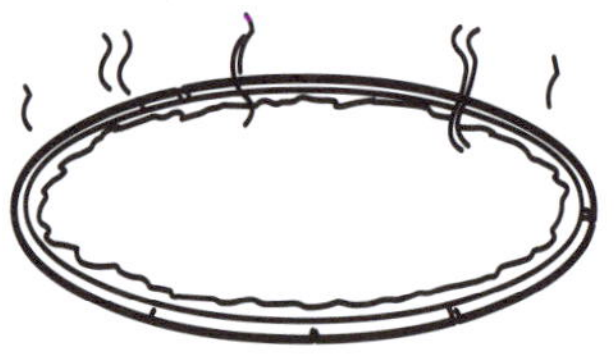 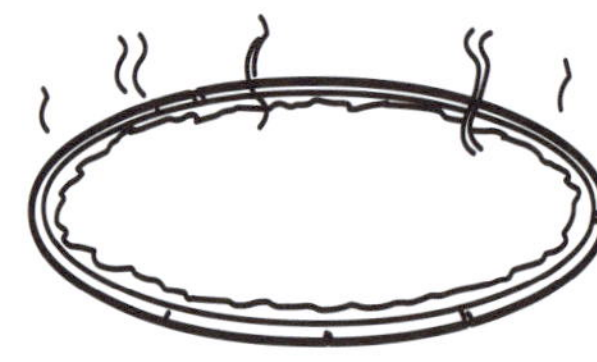 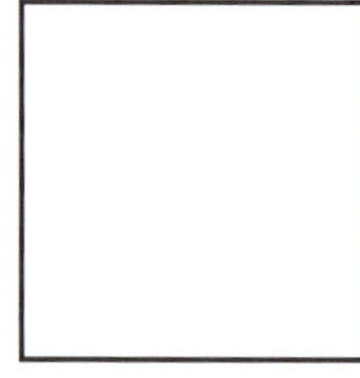 + 6 = 12

1. Show your child how the totals will be the same if you add 7 + 6 instead of 6 + 7. This will help them understand addition more clearly.

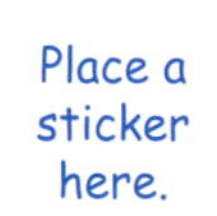

up to 15 or less

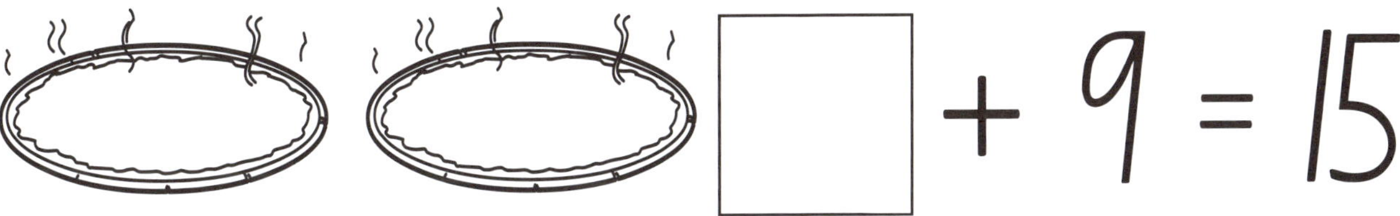

$\square + 9 = 15$

$\square + 4 = 13$

$\square + 10 = 14$

2. Cut out 10 red and 10 blue cardboard shapes. Call out a number up to 15 (eg. 13). Find different ways to make this number using red and blue shapes (eg. 8 blue and 5 red).

Place a sticker here.

Adding up to 20 or less

Draw eggs on the left of the hen to match the first number. Draw more eggs on the other side of the hen to match the second number. How many eggs are there altogether? Write the number in the boxes.

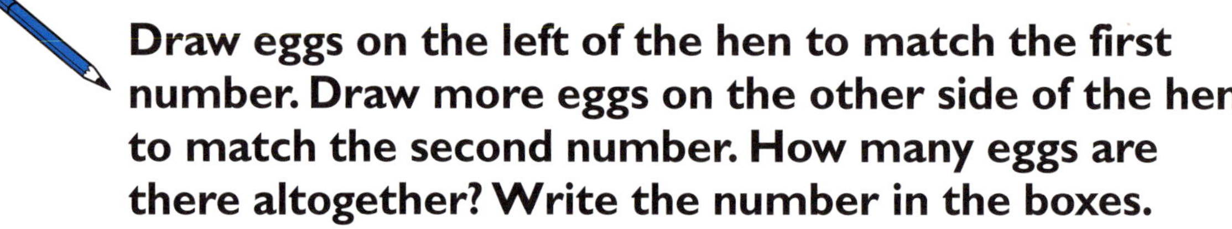

9 + 7 = 16

8 + 9 =

8 + 8 =

1. Find two empty containers. Put some buttons in both. Ask your child to count how many in each container, then how many altogether.

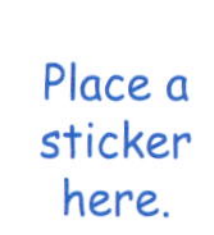

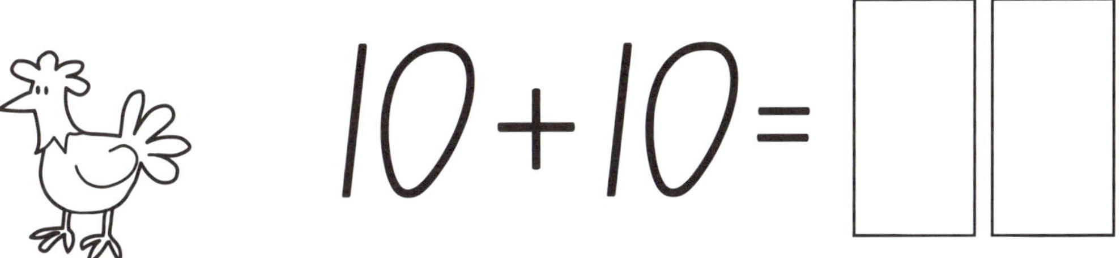

$10 + 10 = \square\square$

$9 + 8 = \square\square$

$7 + 9 = \square\square$

2. You can cut and paste pictures from magazines onto a sheet of paper, then help your child write their own addition sums to match the numbers of pictures.

Place a sticker here.

Adding missing numbers up

Draw spots on the first cow to match the first number. How many spots will you draw on the second cow so that the numbers add up? Try counting on your fingers, then write in the missing number.

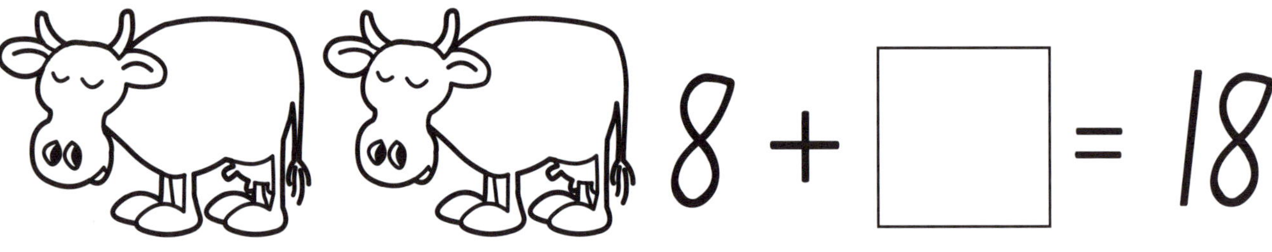

1. Help your child work out each problem using counters (eg. pencils or other small objects).

Place a sticker here.

to 20 or less

$9 + \square = 17$

$8 + \square = 16$

2. Discuss with your child the different ways you can solve each problem. Can they use their fingers? Can they count objects? Can they count in their head?

Place a sticker here.

Adding more missing numbers

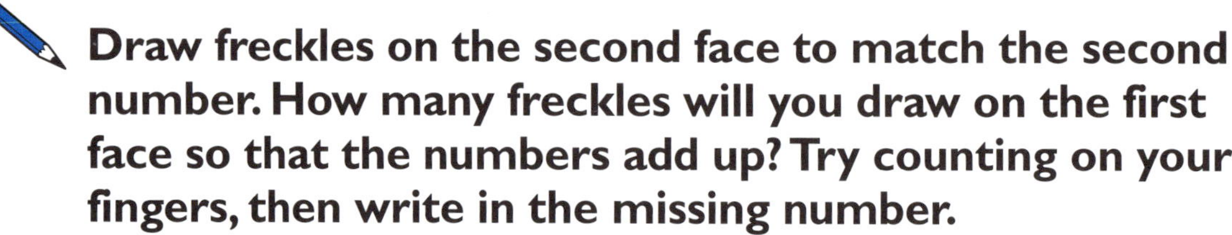

Draw freckles on the second face to match the second number. How many freckles will you draw on the first face so that the numbers add up? Try counting on your fingers, then write in the missing number.

10 + 8 = 18

☐ + 10 = 19

☐ + 7 = 16

1. You can help your child start counting from the second number as they draw freckles on the first face, counting until they reach the total amount.

Place a sticker here.

up to 20 or less

$\square + 9 = 17$

2. Put buttons into 2 piles. Tell your child how many there are altogether. Then cover up one pile and ask your child to guess how many are hidden.

Place a sticker here.

Counting on to 20

Imagine you are a frog. You can jump forwards on a number line to help you add. Start at 0 and jump all the way to the first number. Then count on by small jumps to match the second number. The number you land on is your answer.

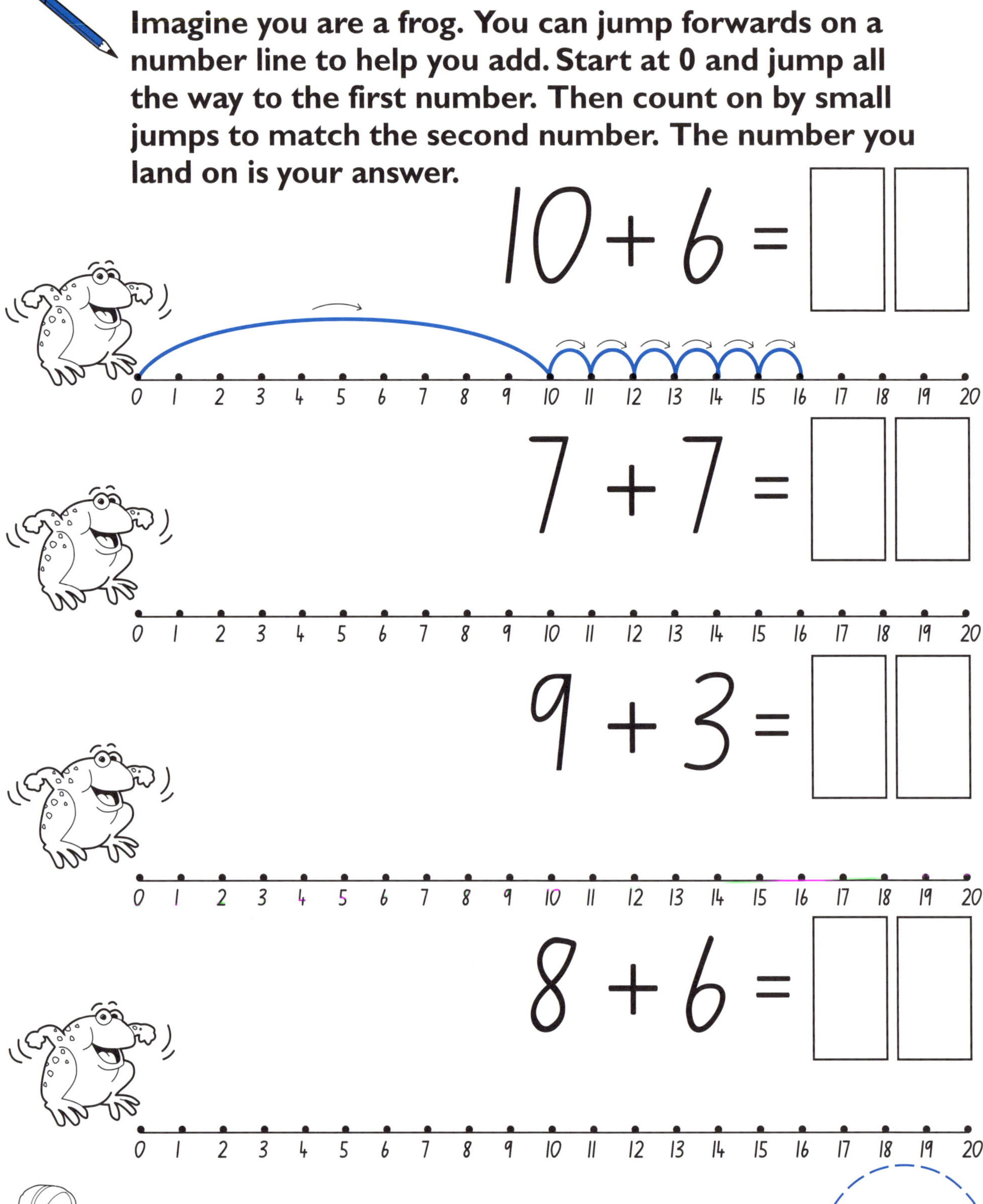

1. Draw a number line outside with chalk and repeat each of these activities with your child jumping to each number.

Place a sticker here.

$10 + 3 = \square\square$

0 1 2 3 4 5 6 7 8 9 10 11 12 13 14 15 16 17 18 19 20

$9 + 5 = \square\square$

0 1 2 3 4 5 6 7 8 9 10 11 12 13 14 15 16 17 18 19 20

$8 + 7 = \square\square$

0 1 2 3 4 5 6 7 8 9 10 11 12 13 14 15 16 17 18 19 20

$9 + 4 = \square\square$

0 1 2 3 4 5 6 7 8 9 10 11 12 13 14 15 16 17 18 19 20

2. Does your child know the answer in their head before completing the number line? Ask them to guess the answer first.

Place a sticker here.

Finding the partner to 20

The numbers on the shirts add to the number in the squares. Use your fingers or a set of counters to find the missing number. Then write it on the blank shirt.

5 7 | 12

8 | 17

10 | 20

1. You can help your child add these numbers by using a set of counters (eg. dried beans).

Place a sticker here.

9 | 18

8 | 13

6 | 14

2. How many different ways can your child put 19 counters into 2 groups? Record your discoveries on a sheet of paper.

Place a sticker here.

Taking away from 10 or less

The first number you see in each row matches the number of faces. Cross out faces in each group to match the second number. Then count how many faces are left and write this in the circle.

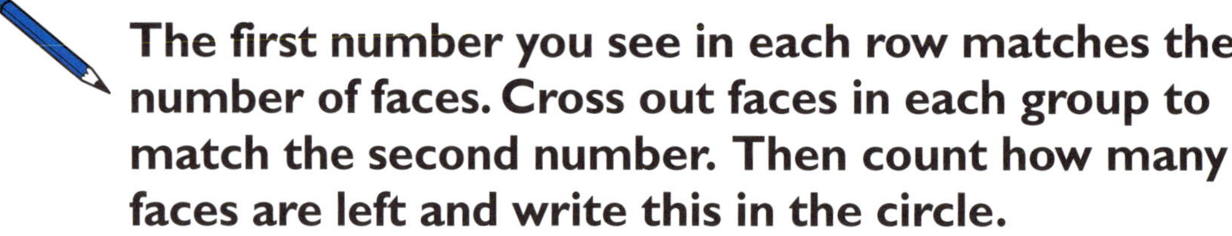

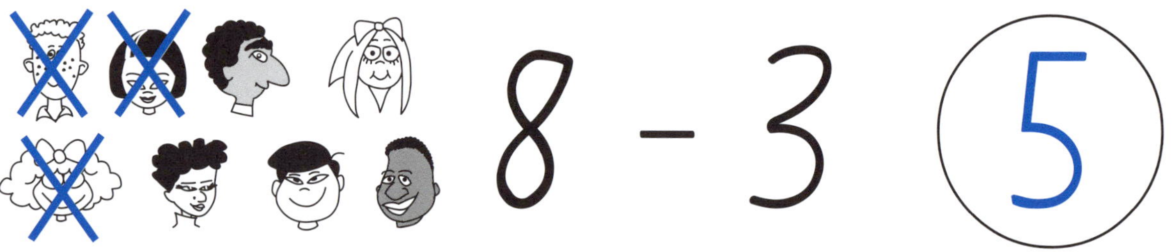

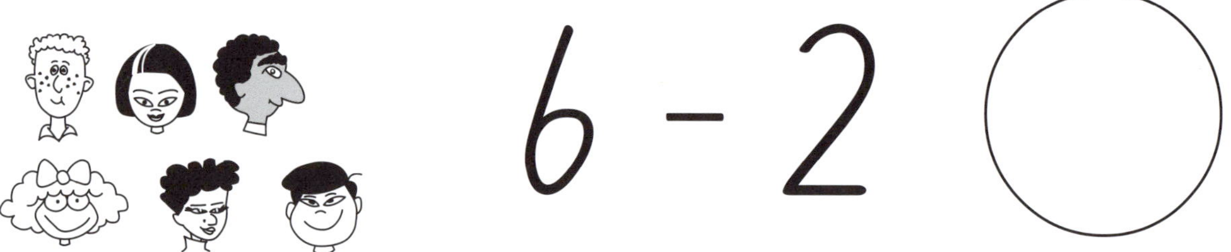

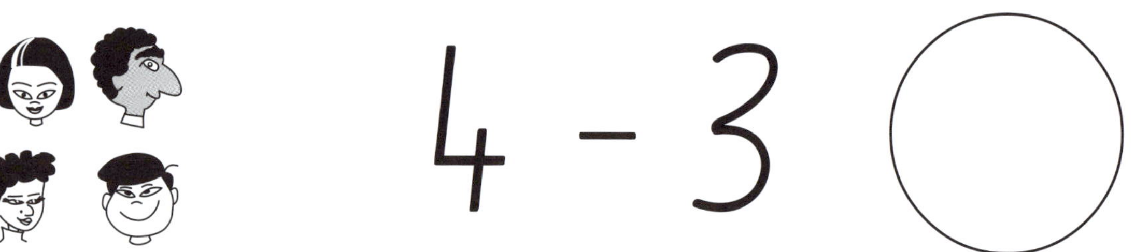

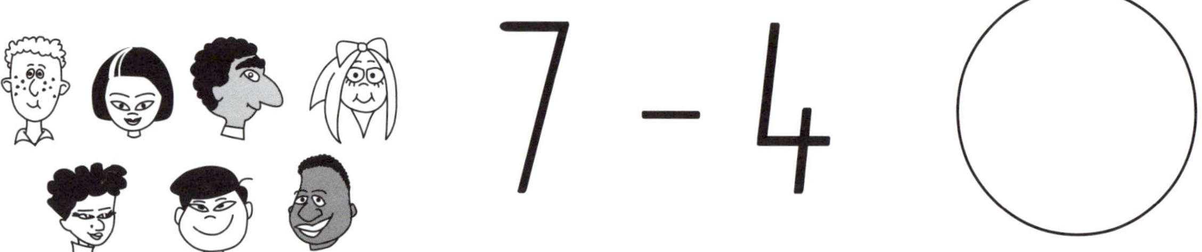

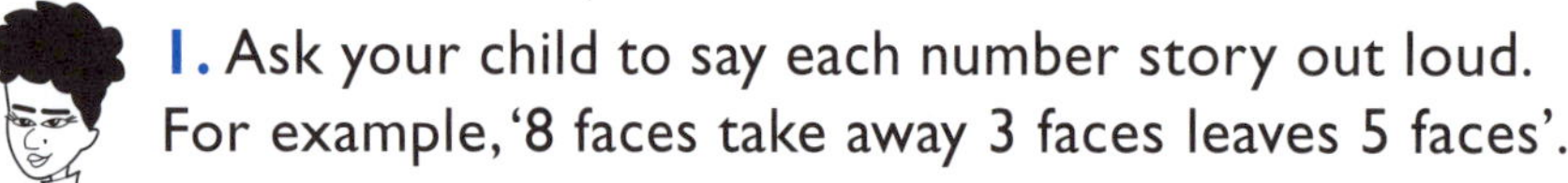

1. Ask your child to say each number story out loud. For example, '8 faces take away 3 faces leaves 5 faces'.

Place a sticker here.

5 - 3

8 - 4

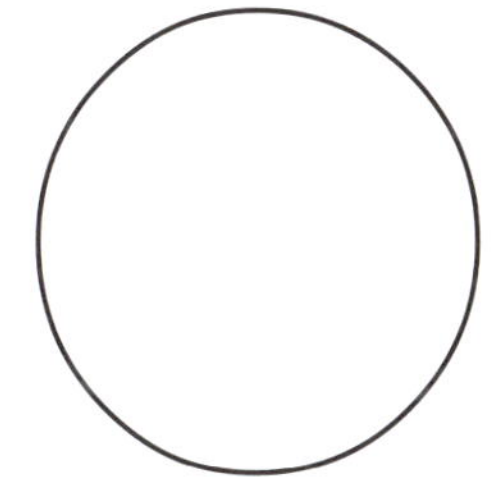

2. Collect a group of soft toys to match the first number. Then put the second number of toys back in the toybox. How many are left each time?

Place a sticker here.

Taking away from 10

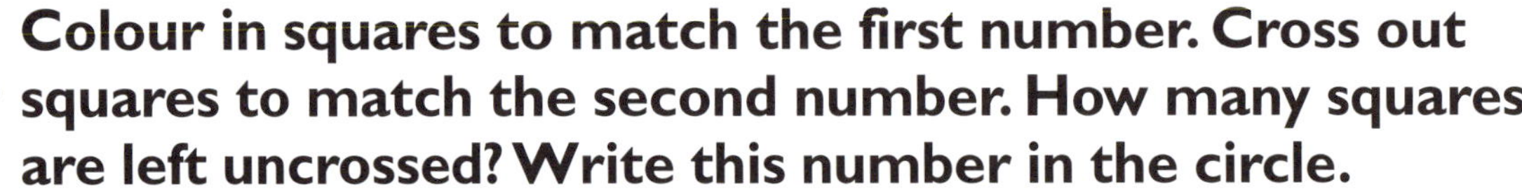

Colour in squares to match the first number. Cross out squares to match the second number. How many squares are left uncrossed? Write this number in the circle.

1. Ask your child to hold up 10 fingers, then fold down 1–10 fingers. How many remain each time?

Place a sticker here.

10 − 7 ◯

10 − 9 ◯

10 − 6 ◯

10 − 8 ◯

2. Find 10 socks. Remove 1–10 socks. How many are left each time?

Place a sticker here.

Counting back from 10

Imagine you are a frog. You can jump on a number line to help you take away. Start by jumping from 0 to your first number. Then jump back in small jumps to match the second number. The number you land on is your answer.

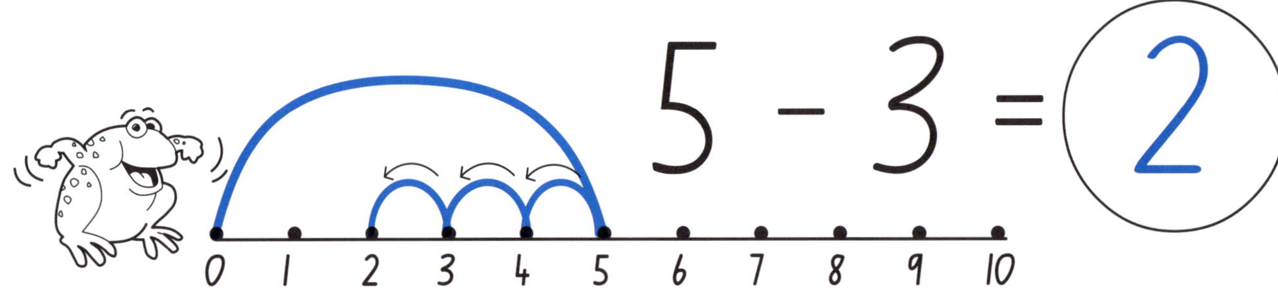

4 - 3 =

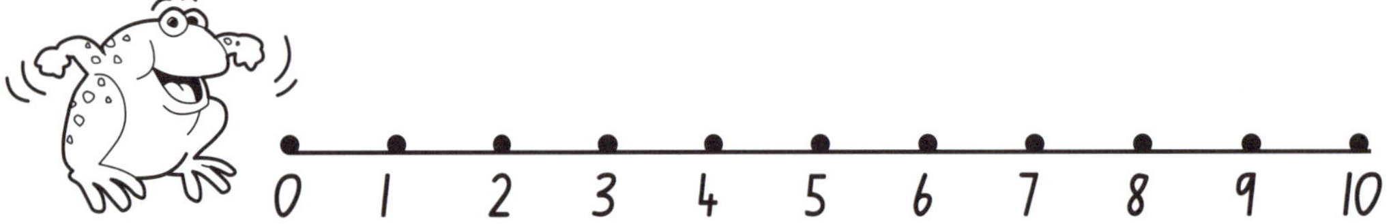

6 - 2 =

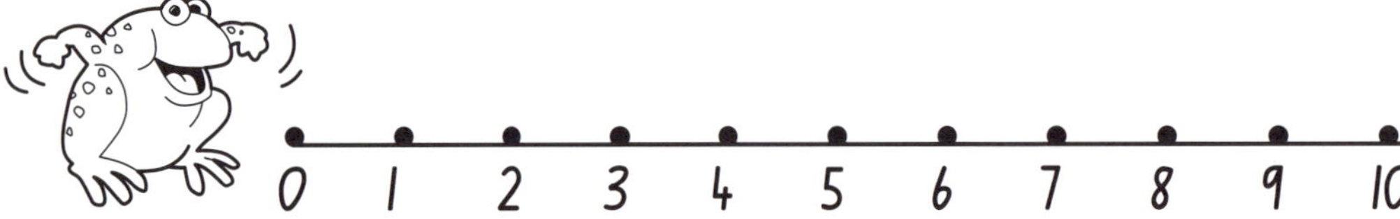

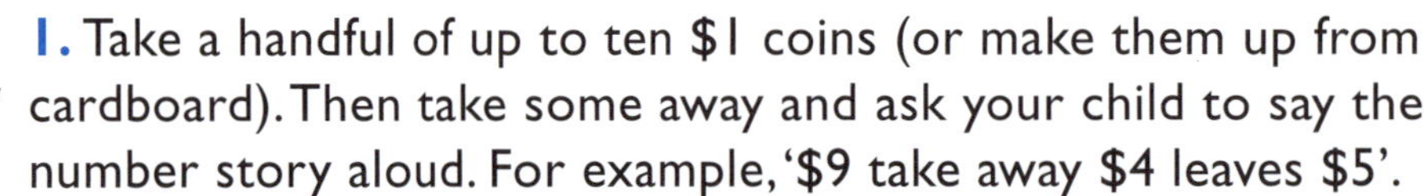

1. Take a handful of up to ten $1 coins (or make them up from cardboard). Then take some away and ask your child to say the number story aloud. For example, '$9 take away $4 leaves $5'.

Place a sticker here.

$9 - 4 = \bigcirc$

0 1 2 3 4 5 6 7 8 9 10

$7 - 3 = \bigcirc$

0 1 2 3 4 5 6 7 8 9 10

$8 - 2 = \bigcirc$

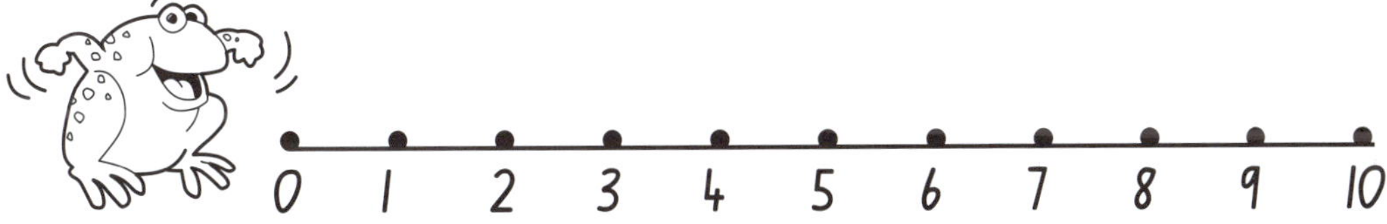

2. Draw a number line on a chalkboard or whiteboard surface and ask your child to draw in a subtraction like the ones on this page. Then rub it out and do another one.

Place a sticker here.

Taking away from 15 or less

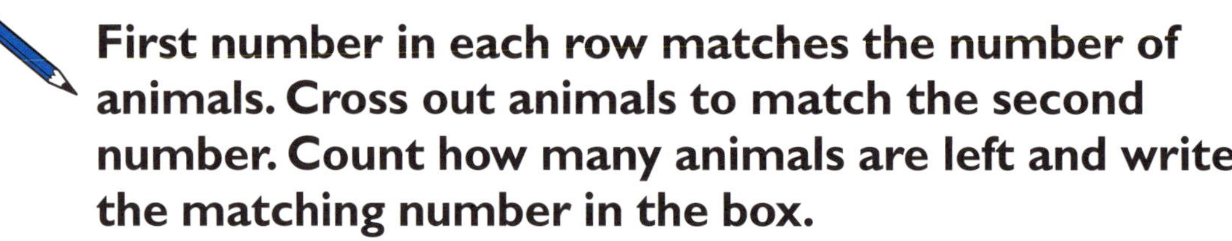

First number in each row matches the number of animals. Cross out animals to match the second number. Count how many animals are left and write the matching number in the box.

1. Ask your child to check that there is the correct number of pictures in each group before crossing any out.

Place a sticker here.

2. Take 2 empty containers and 15 buttons and use these to match each problem on the page.

Place a sticker here.

Taking missing numbers away

How many dinosaurs do you need to take away so the number remaining matches the number on the right? Count backwards and cross out one dinosaur at a time until you reach the right number, then write the number you have crossed out in the box.

13 − [6] = 7

11 − [ ] = 5

10 − [ ] = 8

1. Ask your child to check that there is the correct number of dinosaurs in each group before crossing any out.

Place a sticker here.

from 15 or less

$15 - \square = 5$

$13 - \square = 4$

$15 - \square = 8$

2. How many will be left in each group of you only cross out 1 dinosaur? 2 dinosaurs? 3?

Place a sticker here.

Taking away from 20 or less

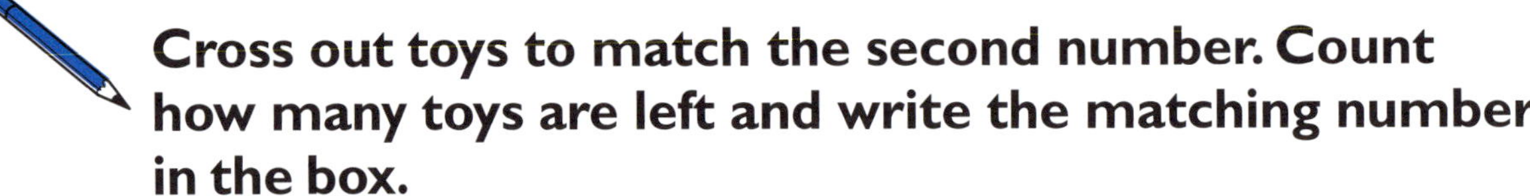

Cross out toys to match the second number. Count how many toys are left and write the matching number in the box.

18 – 10 = 8

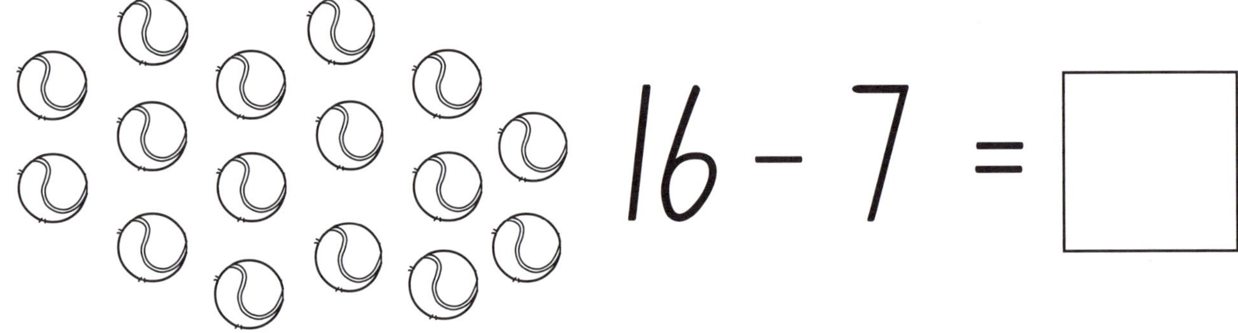

16 – 7 = ☐

19 – 9 = ☐

1. Ask your child to check that there is the correct number of pictures in each group before crossing any out.

Place a sticker here.

$19 - 10 = \square$

$17 - 9 = \square$

$20 - 10 = \square$

2. Ask your child to say each number story out loud. For example, '19 cars take away 10 cars leaves 9 cars'.

Place a sticker here.

Taking missing numbers away

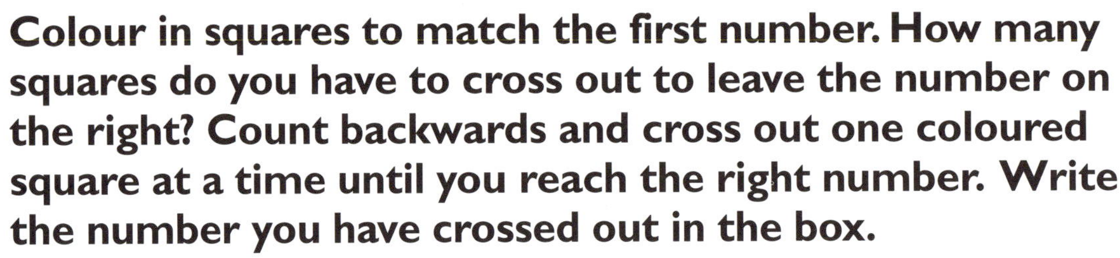

Colour in squares to match the first number. How many squares do you have to cross out to leave the number on the right? Count backwards and cross out one coloured square at a time until you reach the right number. Write the number you have crossed out in the box.

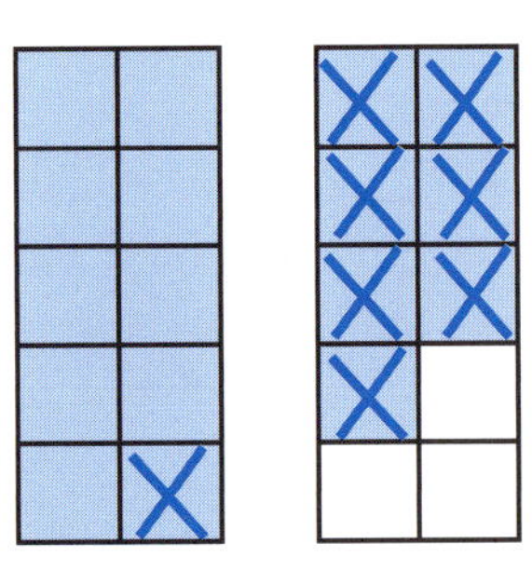

17 − 8 = 9

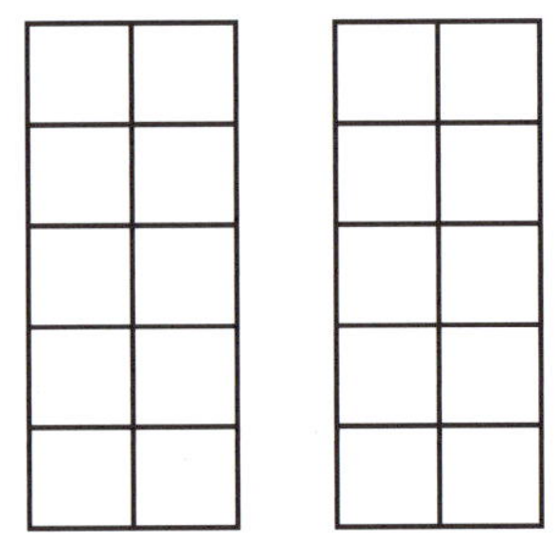

16 − ☐ = 9

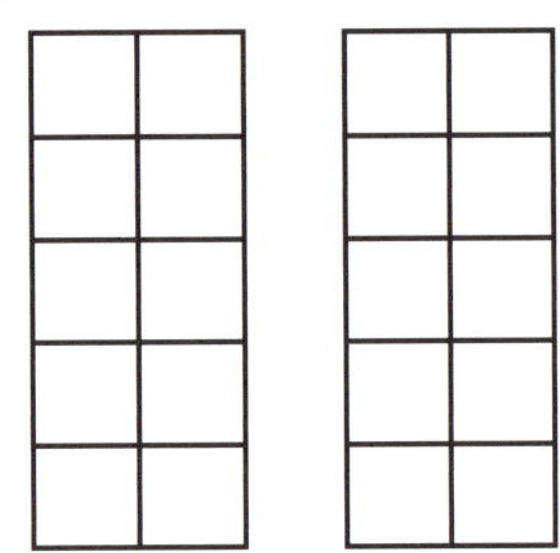

18 − ☐ = 9

19 − ☐ = 9

1. Practise counting backwards from 20 with your child. Ask your child to jump or step backwards when they say each number.

Place a sticker here.

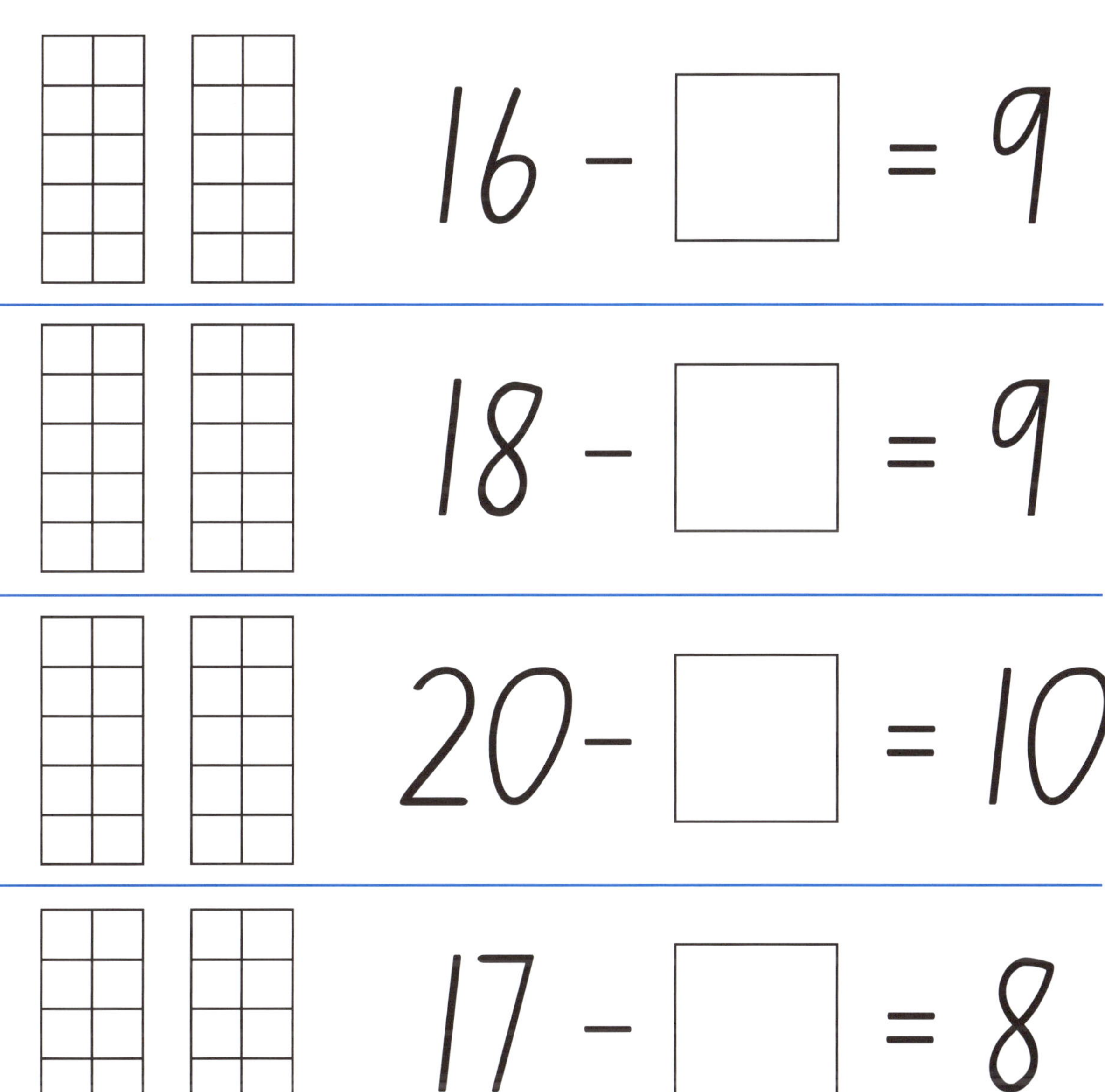

2. If your child is confident with these activities, you could show them how the last two numbers will add up to make the first (eg. 16 – 8 = 9, and 8 + 9 = 16).

Place a sticker here.

Counting back from 20

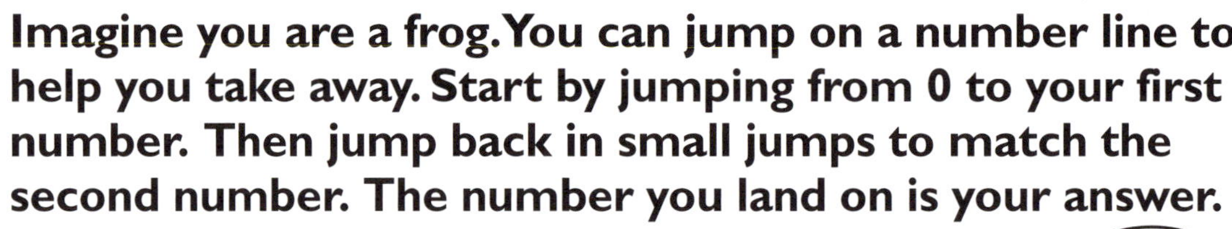

Imagine you are a frog. You can jump on a number line to help you take away. Start by jumping from 0 to your first number. Then jump back in small jumps to match the second number. The number you land on is your answer.

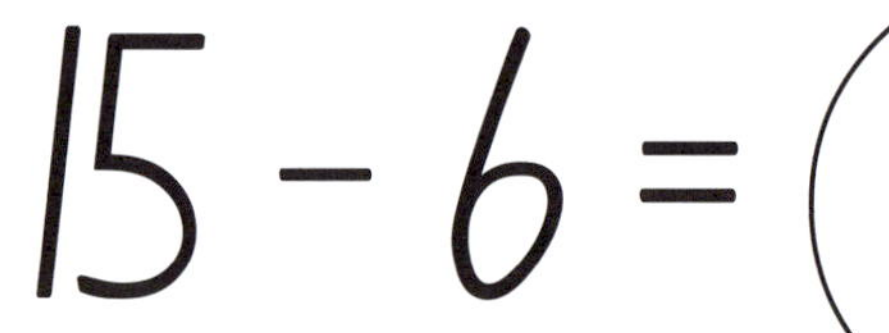

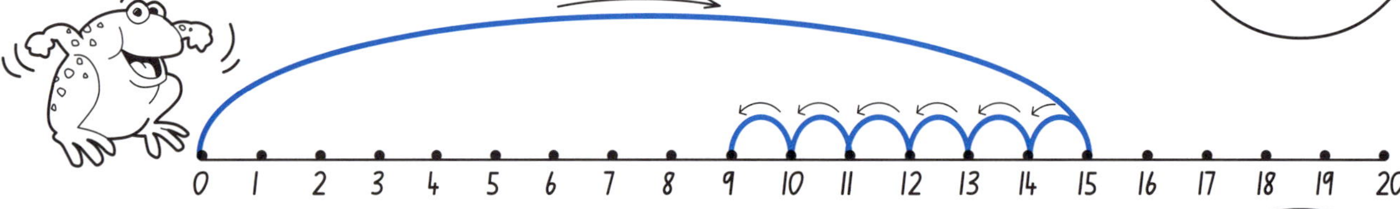

13 - 7 =

0 1 2 3 4 5 6 7 8 9 10 11 12 13 14 15 16 17 18 19 20

11 - 8 =

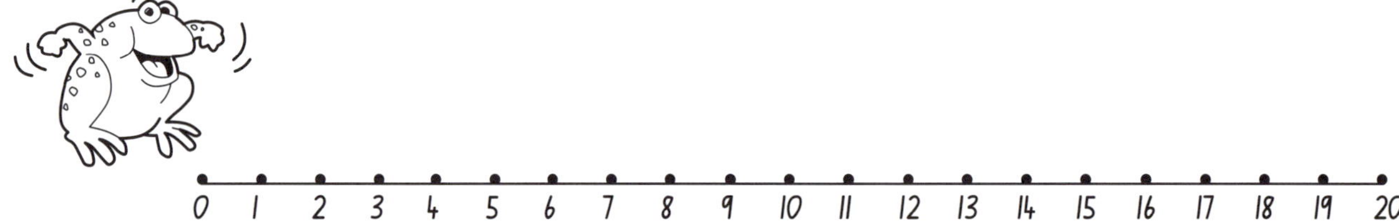

1. Ask your child to make up a number story for each problem on the page. For example, '15 frogs are on a log, and 6 jump off. There are 9 frogs left.'

Place a sticker here.

$16 - 6 =$ ◯

0 1 2 3 4 5 6 7 8 9 10 11 12 13 14 15 16 17 18 19 20

$18 - 9 =$ ◯

0 1 2 3 4 5 6 7 8 9 10 11 12 13 14 15 16 17 18 19 20

$17 - 4 =$ ◯

0 1 2 3 4 5 6 7 8 9 10 11 12 13 14 15 16 17 18 19 20

2. Use a set of counters to match each story. For example, use 16 counters, then take 6 away. How many are left?

Place a sticker here.

Well done!

You have finished the book!

Place your last two stickers on the picture.
You can colour in the picture, too.